***

THE UNIVERSAL THINKER
SCIENCE, QUESTIONS, AND THE UNIVERSE

Copyright © 2010 by Feilong Wu
All rights reserved

Printed by CreateSpace

Cover Design by Megan Steward

ISBN 978-1450562546

THE UNIVERSAL THINKER

SCIENCE, QUESTIONS,
AND THE UNIVERSE

*

Feilong Wu

Acknowledgments

I could acknowledge my thankfulness with pages half the length of this book. To my parents, who are the most direct link to my humble existence; to my mentors, who have influenced me throughout the years; to my wife, who is unquestionably the better half of me; to my friends, who offer me unconditional supports; and to strangers, who I can always learn from…

However, ***it's the thought that counts*** that saves me the pages. Beyond my immeasurable appreciation for life, planet Earth, and the Universe, above all, I'm thankful for thought. And it's thought that makes everything possible.

Thinking about the Universe?

Keep thinking.

Contents

Preface		9
Introduction		11
?.	???	13
1.	The Beginning	15
2.	The Universe	21
3.	Space	29
4.	Time	37
5.	Light	49
6.	Senses	57
7.	Temperature	63
8.	The End	69
9.	Matter	79
10.	Reality	89
11.	Numbers	99
12.	Mathematics	107
13.	Life	117

14.	Intelligence	131
15.	Knowledge	141
16.	The Mind	149

Preface

I don't think I know but I know I think. ***The Universal Thinker*** is about thinking; I think. I have been thinking about writing this book for quite some time now, and perhaps I am not thinking straight for actually doing it.

Here is a question: How can a person like me who doesn't have much formal education, knows very little about science and philosophy, and as a matter of fact, doesn't know much about anything, write a book about seemingly lofty subjects? The answer: One thing I know is that we have to be thinkers before we become scientists, philosophers, and anything else. It is doubtful that I know a small fraction of that of a learned person, but I am certain that I am as much a thinker as anyone else. We must always think regardless what we know or don't know. If thinking is universal then I should have no fear getting into subjects that are derived from thinking.

The next question: Why would I, a person who speaks relatively limited English, is lousy in grammar, and can hardly spell words, want to write anything in this mind-baffling language (that's what THEY say about English); especially on material that is supposed to be somewhat thought-provoking? I have no sound answer to this one; but I think that thinking requires no language. Thinking gives birth to ideas, and knowledge provides us with tools to express the ideas. Since thought comes before knowledge and ideas precede words, and if my way of thinking is anywhere near being right, then, I should have no trouble writing something as a result of thinking, even if I am doing so in my second language.

Evidently, I am neither a scientist nor a philosopher; and I am far from being a competent writer. So with what have I written a book about the Universe and thinking? I wrote it with thought, and thought is all I have to offer. What is my thought? Answer: We are what we think. I think.

Introduction

How big is the Universe? Where did the Universe come from? When will the Universe end? Who are we here to observe the Universe? Why do we think about the Universe? What do we know about the Universe? Countless questions about the Universe have been raised repeatedly by mankind throughout the ages; and answers to our own questions have been continuously refined, re-thought, and re-challenged. For centuries, scientists have been a main force in shaping our views of understanding the Universe. And for millennia, philosophers have been a central guide to our quest for wisdom of the Universe. While scientific advances provide men with forefront knowledge, philosophical thinking continues to redefine those who possess the knowledge. What we know influences what we think, and vise versa.

The desire to know is in the heart of every man; and the struggle to understand is in the mind of every thinker. It is under this inherent human condition, accompanied by an inseparable companion, The Universal Thinker, or The Thinker for short, contemplates fundamental concepts concerning the conditions of the Universe including space, time, matter, and most of all, us. Some answers have been provided throughout the discussion. However, there are no final questions and answers; because every question could be answered, and every answer could be questioned. In this little and finite book, questioning the unquestionable and answering the unanswerable, The Thinker presents us with a Universe of infinite thoughts. Do we really know, or do we think we know? For as long as you are thinking while reading ***The Universal Thinker***, you will become your own universal thinker.

?

???

Any questions?

When?	Now.
Where?	The Universe.
Who?	The Universal Thinker and a companion.
What?	Anything.
Why?	Just because.
How?	Let us find out.

1

The Beginning

•

Let the beginning begin.

It all begins with thinking, thinking about the Universe, before the future, before the present, before the past, and even before the beginning. Thought is never lost; and The Thinker is not alone, for there is always a companion.

The Thinker: It is the brightest and the darkest; it is the farthest and the closest; and it is the known and the unknown. It is the Universe. Does the Universe have a beginning? What does it mean by the beginning anyway?

Companion: The best scientific knowledge mankind has to offer in the 21st century is centered on the *Big Bang Theory* regarding the origin of the Universe. The theory states that some 14 billion years ago, from nothing, or from an infinitesimally small point, an explosion started the birth of the Universe and it has been expanding ever since.

The Thinker: From nothing and we get the whole Universe? Is it not true that we are not supposed to get something from nothing?

Companion: Well, nothing might be equivalent to everything and everything could account for nothing. Before the so-called nothing, there might have been something, and that something became everything.

The Thinker: If there was already something, then it could not be the beginning. Why do we call it the beginning?

Companion: We prefer having a beginning for everything just like we need a starting point for each day. For example, six o'clock might be the beginning of your day. It can be a moment of getting up in the morning, a moment of sunrise, or a moment of some happening, all from your point of view. This moment is very imprecise; but it is a moment you KNOW, and it is a moment you NEED, and to you, this moment is the beginning.

The Thinker: You mean the *Big Bang* is the moment we know, and it is the moment we need, so we call it the beginning?

Companion: To be more precise, the imprecise moment we THINK we know is the moment after the Big Bang. Scientific observations and mathematical calculations lead us to a very small fraction of the first second of the Universe. We do not know what happened at *time zero*, and we cannot define anything before time zero.

The Thinker: This is mind boggling. Let me see, the existence of the Universe came into being right after the Big Bang, and before that, there was no time, no space, no matter, and nothing at all!

Companion: We can only speculate. Amazingly, when it comes to conquering the unknown, the human mind plays the Universe's most fearless warrior. We can speculate that there might have been something and some moment before the so-called big bang moment. But what could have been there before time zero? You see, even when we cannot know any further, we can still ask ourselves: Could there be time before the time we know? Could there be something before anything we can define?

The Thinker: It seems to me that we know of the beginning by what we do not know; and we define the Universe by what we cannot define.

Companion: That is right. However, regardless the conditions of the Universe, we will always want to know and we will always strive to define. So, from what we can define what we know, the Universe emerges from our definition. For the beginning, we start our day at a certain moment, a moment of awareness, and the day unfolds according to our concerns. Let it begin, for this is only the beginning.

2
The Universe

Think about the unthinkable.
Imagine the unimaginable.

The Thinker: It is difficult to fathom the immenseness of the Universe. What is the Universe? How big is the Universe?

Companion: We live in an almost unimaginably vast Universe. The Earth is about 8,000 miles across and it is big enough to accommodate mankind in the billions. The Sun's diameter is nearly 100 times the Earth's size, and the Sun is just one of hundreds of billions of stars that make up the Milky Way galaxy. The Milky Way galaxy is among billions of galaxies that make up clusters and super-clusters…

The Thinker: The Universe is filled with huge clumps of stuff!

Companion: And the distances among all the celestial bodies are enormous. The Sun, our nearest star, is about 93 million miles from the Earth; the next closest stars are over 25 trillion miles, or at least 4 light-years away; the neighboring galaxies and clusters are millions of light-years beyond…

The Thinker: Such a grand Universe!

Companion: With so much material and so much more space in between, the Universe contains a seemingly unthinkable amount of content, yet it is mostly empty. If the Universe has been expanding nonstop ever since the proposed beginning, the Universe is getting emptier as space widens.

The Thinker: This indicates that no matter how big the Universe is it must have a boundary. It simply came from a smaller size, however small it was, to become a larger size, however large it will be. This might sound weird; the Universe is like a balloon that is being blown up! Is the Universe a gigantic balloon?

Companion: Kind of, at least to some people. Earthly imagery makes it a little easier for us to depict the untouchable figure of the Universe, and we attempt to gain understandings of the Universe by giving it some shape or form.

The Thinker: What kind of shape or form, the kind of a balloon?

Companion: More or less. When we observe nature, we see all sorts of shapes, and the most natural ones are those with the best symmetry.

The Thinker: Symmetry?

Companion: Yes, symmetry. A cube, for example, is quite symmetrical. Its top and bottom, left and right, and front and back are virtually indistinguishable. And its change of positions is revealed only by minor features like corners and edges.

The Thinker: Well, a beach ball is even more symmetrical. Whichever way we look at it, it is always round-shaped.

Companion: Right. Objects with spherical shapes have the best symmetry; and a sphere is one of the most natural, most essential, and most beautiful shapes in the Universe. It is no wonder that most significant celestial objects are spheres, such as stars, planets, and moons.

The Thinker: I cannot think of anything that is more natural, more essential, and more beautiful than the Universe itself. The Universe must have a shape with the best symmetry possible!

Companion: Without a doubt; the Universe must have the best symmetry.

The Thinker: But what kind of shape does the Universe have?

Companion: Well, some people are convinced that the Universe has a spherical shape, and it is for good reasons. As mentioned before, a sphere has the best symmetry. Besides, if the Universe expands in all directions at the same rate, the image of an inflating balloon, which is a spherical shape, is a top candidate for the shape of the Universe. Moreover, a sphere does not seem to be bound by any edge. Take the Earth as an analogy; you can head out in any direction on the surface and you will never meet the edge of the world. You might circle back to where you started and go around the globe indefinitely, and you will never leave the Earth.

The Thinker: This is clever. We could head out in any direction in a spherical Universe and we would never meet any edge. Having such shape would give the Universe a finite size with boundary nowhere to be found.

Companion: The idea of a spherical Universe is charming, is it not?

The Thinker: Charming or not, a spherical shape would stipulate the Universe to be finite. A sphere is a sphere no matter how big it is, and it could be the whole Universe? Somehow, it does not sound right!

Companion: Mmmm…

The Thinker: What about the stuff outside the balloon? I mean what about whatever there might be outside the Universe. Does the Universe not include *EVERYTHING*?

Companion: Of course the Universe has to include everything; everything we can think of, including space, time, the tangible, and the intangible.

The Thinker: Let me get this straight. There could be something outside the Universe that we do not take into account, because we cannot think of what that something might be? What if there is another universe outside our Universe, like another star system outside the Solar System? For as much as we *do not* know, there could be countless universes just like there are countless stars out there.

Companion: And countless guesses. Let there be universes outside and inside our Universe, universes before and after our Universe, and universes in every possible way.

The Thinker: Suppose we included all the countless universes and counted everything as one, could we conclude a true Universe that would satisfy the meaning of everything?

Companion: Universally impossible! If any quantity could be included then there ought to be some other quantities outside *The Included;* if any number could be counted then there had to be some other numbers besides *The Counted;* and if any result could be concluded then there must be some other results apart from *The Concluded*.

The Thinker: What are you saying? Are you implying that the Universe cannot be limited, cannot be calculated, and cannot be understood?

Companion: The Universe is bound by the shapes, forms, and calculations of our ordinary cognition, but our inchoate understanding of the Universe should not restrict our imagination. Once we think outside the boundary, the Universe becomes boundary-less. For instance, the Universe requires no space…

The Thinker: What?

3

Space

Let's make room for space.

Companion: The Universe requires no space. We are the ones who demand it.

The Thinker: How?

Companion: Think about it. Why is there space? What is space?

The Thinker: Space is the emptiness between all objects in the Universe.

Companion: What is emptiness? What are all the objects in the Universe?

The Thinker: I am not sure about emptiness. But all the physical objects in the Universe are made of matter.

Companion: Speaking of matter, let us consider it for a moment before the subject of space. We can loosely define matter as anything that has mass and occupies space. We might want to refer to matter as ordinary or normal matter, because we also have names for other forms of matter, such as *Antimatter* and *Dark matter*.

The Thinker: What is the difference among all matters?

Companion: At the microscopic level of ordinary matter, the atoms are composed of positively charged protons and neutral neutrons in the nuclei, and they are surrounded by negatively charged electrons. We exist in the form of ordinary matter. In contrast, Antimatter simply has an opposite charge than ordinary matter. It has been said that in the early Universe, energy created particles in pairs, namely, particles and antiparticles. Somehow, in the process of forming matter and Antimatter, particles outnumbered antiparticles in the cosmos, and matter ended up winning the battle of survival. We would not exist if there were more Antimatter than matter, equal amounts of matter and Antimatter would annihilate each other, and all things would disappear in the air (although there would not be air either). As to Dark matter, it refers to the huge amount of unseen matter—multiple times more than ordinary matter—that has been calculated to exist around galaxies and clusters. It is *dark* and invisible because it does not emit light like ordinary matter. Its theoretical existence has only been indirectly confirmed, and so it is said to be *missing*...

The Thinker: So complicated!

Companion: Exactly! We are generally in the dark about things we do not see, and you might say that there has to be more ordinary matter than Antimatter, because if there were more Antimatter than ordinary matter, and we were still here to make evaluations, we would see Antimatter as the normal form of the Universe. And naturally, the so-called Antimatter would just be ordinary matter, and the so-called ordinary matter, as we recognize now, would just become Antimatter...

The Thinker: Hey, you are trying to make the matter even more complicated.

Companion: Quite contrary. The point is that it does not make any difference if there is ordinary matter or Antimatter, visible matter or unseen Dark matter, or any other types of exotic matter, all are physical stuff of the Universe, and can simply be called matter.

The Thinker: OK. Let matter be matter. What about space? What does matter have anything to do with space?

Companion: Matter has EVERYTHING to do with space! How do we know the emptiness in the Universe? We know it by the contrast of the presence of matter. And we obtain space from measuring the distance between physical objects. Without the Sun and the Earth, there would not be millions of miles of space; without stars, galaxies, and clusters, there would not be millions of light-years of space; and without matter, there would be NOTHING in the Universe.

The Thinker: Nothing? What about *light, force fields, energy,* and *Dark energy?* What are they? They seem to be anything but matter.

Companion: They are matter, or effects of matter. Matter converts into energy, creates forces, and emits light. As to Dark energy, which we can neither see nor detect, it is speculated to exist in all of space to explain the seemingly accelerating expansion of the Universe. Speculation or not, if space everywhere is filled with Dark energy then there is no such thing as empty space, not even in a vacuum. And if it really physically exists, it is part of matter, or at the least, a manifestation of matter.

The Thinker: Nothing else but matter! Well, it is easy to understand that without matter we could not know space. But can we also say that without space we could not know matter? Logically, space and matter need to coexist to support the existence of each other.

Companion: True, logically true! And this is our problem. Our earthly logic confines us to think of the Universe as a gigantic container, and matter is contained in the space within it. We are quick to envision our presence in confined space, but can we picture our existence in pure logic? Can the Universe be contained by our notion of space? Does the Universe exist according to our logic?

Well, with or without logic, we physically exist in the form of matter. We are matter, not emptiness, and *matter is the Universe as we know it.*

The Thinker: Aha! If the Universe requires no space then it cannot be bound by space. It is space-less and it is boundary-less.

Companion: Unlike the Universe, we are bound by the space we need. Space exists in the imagination, and our imagination in itself is imaginary.

The Thinker: Let space be imaginary. Now what about the age of the Universe? Is it not true that the Big Bang Theory's spatial expansion gives the Universe the 14 billion years or so of age?

Companion: Actually, our astronomical observation of an expanding Universe is one of several key points leading to the general acceptance of the Big Bang Theory. From scientific studies, galaxies are *seen* to move away from one another, and space is *viewed* as expanding and carrying the galaxies along with it. If we reverse the process of the observation we can interpret and calculate the time it takes an expanding Universe to contract into an infinitely small point, in a state of infinite density and temperature, all the way back to the beginning. And the Big Bang serves an explanation as a likely triggering event of the expansion of the Universe.

The Thinker: We are dealing with a very popular theory here. But if we boldly disregard the theory and dismiss the existence of space, then there is nothing to expand; and if our current observations are erroneous, we can be misled by the apparent age of the Universe, since the time calculated might be totally wrong.

Companion: Observations can be misleading; however, there is no need for time to be wrong, because the Universe requires no time.

The Thinker: What?!

4

Time

Time is what we don't have.
But we never run out of time.

Companion: The Universe requires no time. We are the ones who demand it.

The Thinker: How?

Companion: Well, what is time and why do we have time?

The Thinker: It is hard to say what time is. It is not an object; it is not an event; and it is not even a medium. But we have to have time to distinguish all the events and happenings.

Companion: Right. *We have to have time!* And we attain it according to the motions of physical objects. For example, the *24-hour day* is essentially derived from the Earth's own rotation. The observations of the cyclic movements of the Moon, the Sun, and stars give earthlings the *month, season,* and *year*. Of course those time units are very imprecise, and smaller units of time are needed. The hour unit is divided into the *minute,* and the minute is divided into the *second*. Beyond every day's common use, much smaller units of time are required for scientific work. From second (the base unit, now measured by atomic clocks, is based on a regular signal emitted by electrons changing energy state within an atom) we get subdivisions such as *millisecond,* which is one thousandth of a second, and *nanosecond,* which is one billionth of a second. Still, tinnier units of time can be further given by means of measuring subatomic particles traveling across certain subatomic distance at a certain speed. As you can see, from stars, planets, and from subatomic particles, it does not matter how large or how small the units are, time is calculated from matter.

The Thinker: Now I am getting the idea that we have invented time as a measuring system for sequencing our world, and watches, clocks, and other timepieces are just guardians of this system. But what about times like the *past,* the *present,* and the *future?* They seem so real, they cannot possibly be invented!

Companion: The past, present, and future are not units of time. They are our concepts of time. How do we conceive those concepts? We conceive them by differentiating the changing stages of matter.

Let us use an egg as an analogy. Think of it as the present when the egg is freshly cooked, then we can consider it as the past when the egg was raw, and the future is when the cooked egg becomes rotten. A cooked egg cannot turn back into a raw egg, and we cannot own the past, because we are *not having* the past. We might try to hold on to the freshly cooked egg as the present, but of course the cooked egg is continuously changing its physical state, such as becoming a rotten egg, and we cannot possess the present, because we are *not having* the present. Then is the rotten egg really the future? No! At what exact point do we know the rottenness of the egg? We know it only when it is rotten, not before it becomes rotten, and we cannot occupy the future, because we are *not having* the future. Physically, the egg is always on its course of transformation with or without our notion of time. We treat our temporal experiences as the present; we keep our fading memories as the past; and we retain our forward expectations as the future. Having to keep up with the ever changing nature of the world, we naturally conceive the concepts of the past, the present, and the future. They are among our most natural inventions.

The Thinker: What about *the arrow of time* we often talk about, is it a mistake? And what about our *lifetime* that sounds so organic, is it just an illusion?

Companion: The arrow of time refers to the flow of time toward one direction without reversibility. What happens is not the flow of time, but only the evolving nature of matter. The arrow-like and irreversible direction applies to matter, such as an egg, not time. We simply use our invention of time to describe the manifestation of the change of matter. Similarly, our lifetime is dictated by the life cycles of organic matter, such as the divisions of cells and regenerations of components that make up the body. What is organic is life and matter, not time. Matter is in disguise, and it creates the illusion of time.

The Thinker: Ah! If the Universe requires no time then it cannot be limited by time. It is timeless and it is limitless.

Companion: Unlike the Universe, we are bound to be limited by the time we need. The existence of time lasts as long as the duration of our concept of time.

The Thinker: Let time be conceptual. Let us dismiss space and time altogether. Now what do we make out of our world? Do we not live in a world with three dimensions of space and one dimension of time?

Companion: We live in a dimension of our own; and our notion of the dimensions of space and time exemplifies the peculiarity of our creative dimension of learning. We acquire the dimensions and we complicate the world by our own complication.

The Thinker: What do you mean?

Companion: To describe our presence, we learn to orient ourselves on three independent directions, namely, *left and right*, *back and forth*, and *up and down*. How do we come up with such invisible orientation? We do it via the presence of physical substance. And the presence of physical objects is commonly located by the three familiar terms we use to measure our worldly objects; and they are *length*, *width,* and *height*. Inevitably, from such measurements of physical presence, we concede that the three dimensions must be the way of our space.

The Thinker: Are you saying that we are imposing the three spatial dimensions on our world? Do you not see that space with three dimensions is such a familiar and natural thing?

Companion: *Thing* is right! We can impose forms on formless reality and we can make nothing into something. Strangely, we do it naturally.

The Thinker: False or not, I am quite comfortable with spatial dimensions; since we can use them to locate *every point* in space. Time as a dimension, on the other hand, always strikes me as odd. I suppose we can find *every event* in time, but why the dimension?

Companion: The dimension of time is odd. Here is something about time: First we learn that time is constant, meaning *it is the same* for everyone everywhere at all time. Then we learn that time is not constant, and *it is not the same* for all the observers, depending on their relative traveling speed.

The Thinker: I have gotten the first part. What is the second part about time again?

Companion: Time is relative to the speed you move in space. Simply put; the faster you go the slower time becomes, and if you could travel at the speed of light, time would stop for you, in comparison to the regular time for everyone else.

The Thinker: Marvelous!

Companion: Marvelous but plain and simple; we have learned to acknowledge a correlation between space and time, they are not absolute, and they are dependent on each other. As a result, we have accepted the four space-time dimensions to be the way of our world.

The Thinker: It is difficult not to accept dimensions, especially those with space. Are we getting too much ahead of our learning?

Companion: Our learning takes us to great height; but learning can be arbitrary. We view and measure the Universe in ways with our increasing knowledge. We learn to recognize space through the presence of physical objects; and we conceive time through the motions of physical objects. Space and time seem to be undeniable constituents of the Universe. However, without matter, there would not be anything to view and to measure; there would not be any direction to pursue; and there would not be any dimension to speak of. We occupy the presence of space with matter; we follow the rhythm of time with matter; and there are as many dimensions as the lines that we are able to and willing to draw from our world.

The Thinker: A peculiar world we are in! So what about the fantastic ideas of manipulating *time travel* and space, *the fabric of the Universe?* Are they wishful fantasies of our imagination?

Companion: They are theoretical realities of science and entertainment plays from science fiction. When space is treated like physical fabric, it could warp, curve, and bend; and shortcuts could possibly be created to reach far regions of the Universe. When time is compared with the relative motions and views of different observers, it could speed up or slow down; and traveling to the past and future might be achieved. Indeed, they are fantastic ideas. But if there were space fabric to manipulate, the Universe would be disorderly. If there were different rates of time to temper with, the Universe would be chaotic. No doubt we can always find chaos and disorder in our world, but they are originated from our imperfect observations of the ever changing conditions of the Universe, and for the most part, they are from the conditions of the creative human mind. As apposed to the fabrication of space, matter is the fabric of the Universe, and we travel along with it.

The Thinker: If the Universe cannot be defined by space and time, and matter is all we have; is the Universe as simple as matter?

Companion: Not really.

The Thinker: What else can the Universe be?!

Companion: Before we go any further to examine matter and the Universe; there are some topics we should consider.

The Thinker: Like what?

Companion: Like light and a few other things.

5
Light

We see light in the dark.

The Thinker: I knew it! I knew we could not dismiss light so quickly. Light is such a luminous phenomenon. Let us see what we know about light.

Companion: We know light travels at a very fast speed.

The Thinker: Super fast! In a vacuum, light travels over 186,000 miles per second, or 670 million miles per hour. And in one year, it travels roughly 6 trillion miles, which is the distance of a light-year. Nothing has been measured to travel faster than light. The speed of light is a universal constant and it is the ultimate speed limit of the Universe.

Companion: The claim of light being the speed limit of the Universe needs not to be true.

The Thinker: Why not?

Companion: Let us take a look at the Big Bang Theory again. Remember, we have calculated the age of the Universe to be some 14 billion years; that is, we *see* almost 14 billion light-years in any direction, because it takes light that much time to travel that far. Well, a Universe with that many billion light-years span is quite big, but it is not terribly big that we cannot imagine it.

The Thinker: Right, I can imagine a Universe bigger than that.

Companion: The Universe has to be bigger. The *observable Universe* is estimated to be many times bigger, at least 80 billion light-years across. And there is no reason not to suspect the Universe to be much bigger than our narrow estimation of the Universe.

The Thinker: OK. The Universe is likely to be bigger, so what?

Companion: Do you see the discrepancy between the 80 billion light-years of space and the 14 billion years of expansion of the Universe?

The Thinker: Oh, maybe the force of the Big Bang Theory's explosive bang was so great that it caused the initial rate of expansion of the Universe to be much greater than the speed of light.

Companion: That is one of many proposed theories for solving the problem; a problem that might not exist in the first place. You can see that in order to keep the coherence of the Big Bang Theory, something greater than the speed of light needs to be introduced into the calculation. With such modification, we avoid violating the physical laws of our current understanding, light is still kept as the fastest moving phenomenon *in space*, and it is space itself that is thought to *inflate* at a much faster rate than the speed of light at the very early stage of the Universe. Incidentally, the scenario of rapid expansion of space in the early Universe also solves another problem that arose from our cosmic observations.

The Thinker: And what is that?

Companion: It is called *the Horizon problem*. It has been observed that the Universe has spread out in a way that properties such as radiation and temperature are strikingly uniform everywhere, including far regions of space where information could not have exchanged and shared due to the limitation of the speed of light. But spatial inflation conveniently resolves the conflict. Any way we put it, the speed of light cannot be the speed limit of the Universe, unless…

The Thinker: Unless what?

Companion: Unless of course if you believe in the theory that the Universe is like a hologram.

The Thinker: A hologram?

Companion: Imagine the Universe to be a projected hologram, and far regions of cosmic formations we observe are mere reflections of the closer ones. This way, the Universe appears to be big but it is not big enough to invalidate the speed of light as the speed limit of the Universe.

The Thinker: Interesting! But a hologram seems less convincing than the spatial inflation scenario. Scientific theories can get weird and wild.

Companion: Let us put the theories aside. It is still not hard to see that having the speed of light as a limit puts a limit on the Universe.

The Thinker: What is the limit?

Companion: Light travels fast because we are slow observers. However, light has a very finite speed, and it takes light 80 billion years to travel across just the observable Universe. And in astronomy, we even need units of length that are larger than the light-year for describing the vastness of the Universe. In comparison with the infinite Universe, any finite speeds, including the speed of light, would only seem to be standing still.

The Thinker: We cannot have light as the speed limit, because the Universe has no limit?

Companion: The limit of anything is for us to perceive, and our perceptions often impose limits on ourselves.

The Thinker: Mmmm... that translates into light being the limit only because we see it as a limit.

Companion: And physically, we are limited to how much light we can see. The light the human eye can see is called *visible light* (what else would it be called?), and it has a very small range of wavelengths on the light spectrum.

The Thinker: Wavelengths? Is light like water waves?

Companion: Somewhat; both water and light can bend around corners, so *wavelike* is one way of describing light. Another way to describe light is *particle-like*, because a light-ray's ability of knocking electrons out of metal surfaces suggests the bullet-like property of light.

The Thinker: But the properties of wavelike and particle-like do not mix: Whereas waves flow in a continuous stream, particles flow in discrete packets of energy. How can two totally different properties exist in the same phenomenon?

Companion: Well, that is how we describe the behavior of light. Regardless the true nature of light, our incomplete understanding of nature renders us to say that light is both wavelike and particle-like. Anyway, light can also be called electromagnetic radiation, and on the spectrum we only see it in a tiny range of wavelengths. We cannot see radiation with shorter wavelengths, such as ultraviolet, x-ray, and gamma-ray. We cannot see radiation with longer wavelengths, such as infrared, microwave, and radio waves.

The Thinker: I guess we are limited to see just the visible white light.

Companion: Even the white light we see from the Sun is not white at all. It is only an illusive clear color we see through our sense of sight. We are physically bound by our senses; and we experience the world in ways of what our senses allow us to perceive.

The Thinker: How do our senses perceive?

6
Senses

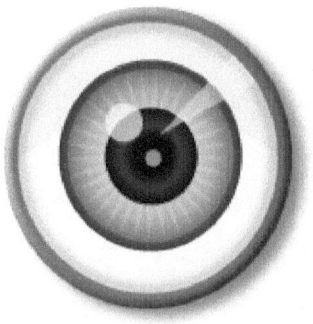

Common sense is common; so is deception.

Companion: We see sunlight as clear white light, but it is made up of many different colors including violet, blue, green, yellow, orange, and red.

The Thinker: It is like a rainbow we see in a rainstorm!

Companion: That is right. Needless to say, those colors are characterized by their wavelengths, or frequencies.

The Thinker: How do we see colors? How do we see what we see?

Companion: The light receptors in the eye detect light, and the brain interprets the detected light as colors according to their wavelengths. Shapes of objects we see are revealed by information gathered from the reflection of light. We see what light allows the eye to see, and we only see what our eyes see.

The Thinker: What about the sense of hearing? How do we hear?

Companion: Hearing is about our perception of sound. Our ears receive the propagating vibrations of sound through a medium like air, and then the vibrations, or frequencies, are reconstructed into sensible sound. We only hear sound with a certain range of frequencies. Frequencies too high or too low are not heard by us. We hear what sound allows the ear to hear, and we only hear what our ears hear.

The Thinker: And the sense of touch, how do we feel?

Companion: Mostly, touch is the perception of pressure on the skin. The skin's nerve endings respond to pressure of different variations, and we feel sensations of various kinds. Pressure too light on the skin we cannot feel and pressure too heavy on the skin will damage the skin's nerve endings and thus deprive the skin of the feeling of touch. We feel what pressure allows the skin to feel, and we only feel what our skin feels.

The Thinker: And what about the senses of taste and smell?

Companion: The sense of taste is a mere result of receptors of the tongue detecting and conveying information to the brain. In a similar fashion, the sense of smell is no more than a process of the receptors of the nose receiving odor molecules and integrating the received information. We taste what flavors allow the tongue to taste, and we smell what odors allow the nose to smell. We only taste what our tongues taste, and we only smell what our noses smell.

The Thinker: Alright! The world might not be what it appears to be, because we perceive the world with our own senses, and our senses can fool our own perceptions. Will it be wrong for me to say that, since we cannot possibly receive the same information and interpret it the same way for vision, sound, touch, taste, and smell, no two persons can see, hear, feel, taste, and smell the world the same?

Companion: No one can tell you that you are wrong when you perceive yourself to be right.

The Thinker: Well, if we cannot truly perceive the Universe in accordance with others, should we not trust our senses?

Companion: Often we have no choice but trusting our senses and our judgments of perception. We rely on and we depend on our senses, however deceptive and false they might be. Besides the five senses we are familiar with, we also have senses of time, of direction, of motion, of balance…

The Thinker: How do we make sense of all those senses?

Companion: Whereas we have to make sense of the world around us, the world does not need to make sense of us. We get to know the Universe through our senses, and the Universe appears to be what our senses perceive.

The Thinker: So we have developed the senses only to be restricted by them, and we have conceived the perceptions only to be constrained by them?

Companion: That is the limitation of our senses and perceptions. Within the limitation, our perception of the sense of temperature is another good example.

The Thinker: The sense of temperature, what about it?

Companion: Talking about the sense of temperature, what does temperature mean to you?

The Thinker: I suppose temperature is the skin's sense of heat or the absence of heat; it is the sensation of feeling hot or cold.

Companion: Naturally, we go by how we feel, and this is common sense. But what is hot or cold?

The Thinker: Fire is hot and ice is cold, respectively. Also we can visualize the hotness and coldness by looking at a thermometer.

Companion: A thermometer is a convenient device we have invented to help us to record temperature readings, and we can record the temperature readings from different temperature measuring scales.

The Thinker: I know the familiar ones are *Celsius, Fahrenheit,* and *Kelvin*. The Kelvin scale is more commonly used by scientists as the absolute temperature scale.

7

Temperature

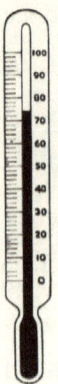

Fire is hot, but is it hot enough?

Companion: Correct; on the absolute temperature scale, *absolute zero* refers to a state of no thermal motion, and it is the coldest temperature possible in the Universe.

The Thinker: That is about negative 273 degrees Celsius, or negative 459 degrees Fahrenheit. Brr… It makes me feel cold just thinking about it!

Companion: Regardless how we feel, temperature is considered not a feeling, but a property of matter.

The Thinker: A property of matter, I think I can see why; since everything must have a temperature. We have 0°C for ice, 20°C for room temperature, 37°C for human, 100°C for boiling water…

Companion: The temperatures we measure and the feelings we acquire are all but an effect from the motions of matter. More precisely, the temperature of everything is determined by microscopic motions of particles, atoms, and molecules of matter.

The Thinker: How does it work?

Companion: Microscopically, restless particles disturb electrons to emit radiation. The faster the particles move, the more radiation is emitted and the higher the temperature can be.

The Thinker: So does it mean a fire we see is just radiation coming out from electrons, corresponding to the wavelengths within the visible light?

Companion: Yes, and as the fire's temperature increases, its color can change from red to orange-red, to white-red, to blue, and eventually it becomes invisible to the naked eye.

The Thinker: Can we continuously increase the temperature of the fire?

Companion: We can increase the fire's temperature to many thousands of degrees, but that is the limit for an ordinary fire that involves *chemical reaction;* that is, the atoms of matter move at a relatively slow speed, and only electrons, not nuclei, are interacted.

The Thinker: What happens if the atoms move fast enough and their nuclei interact?

Companion: Then we have *nuclear reactions,* and temperatures can reach above millions of degrees. That is the temperature of the core of the Sun.

The Thinker: Over millions of degrees, that is hot! It must be destructive. It is hard to imagine being anywhere near such temperature.

Companion: Again, our senses are not in the best position to help us to comprehend the scales of the Universe. The nuclear reactions inside the Sun, or more specifically, the *nuclear fusion reactions* inside the stars, are constructive and important processes. Like numerous other stars, the Sun's high temperature and pressure allow the nuclei to collide and fuse, and generate enough heat and light to last billions of years. It is true that it is hard to imagine what a star's condition is like in temperature of millions of degrees. But try to imagine the Universe without stars like the Sun? You know, without the Sun, there would not be life as we know it.

The Thinker: And of course we would not be here to talk about temperatures, stars, and the Universe. What an unbelievable fact: everything has a certain temperature because everything is moving at a certain speed. We feel temperature by motions, and we see colors by motions.

Companion: And motions still exist even if we do not feel and see.

The Thinker: It is wild. Everything is moving, and we seem to exist in the continuous motions of matter. Does the Universe only exist in motions?

Companion: Motions define the Universe as we know it.

The Thinker: But hold on. Physical substance cannot possibly keep on moving; motions cannot go on forever. What if everything stops moving? Will it be the end of the Universe? Does the Universe have an end??

8

The End

The end is not a dead end.

Companion: Does the Universe have an end? If the end is defined by motions of matter, then the end will eventually replace motions when everything stops. So it all depends on what we mean by saying the end.

The Thinker: What does it mean by us saying the end?

Companion: When we say the end we often talk about *entropy*.

The Thinker: Entropy?

Companion: Entropy is a measure of disorder and randomness of a physical system. Evidently, randomness of the material world relentlessly increases, and the Universe becomes more and more disorderly.

The Thinker: How so? Evidently, this is not very evident to me.

Companion: Evidence demonstrates that macroscopically, heat flows from hotter objects to colder objects, and not the other way around. That is how a glass of hot water cools down in a room, by losing its heat to the environment.

The Thinker: That makes sense. But what about a glass of ice water; how does it lose its coldness?

Companion: Heat from the surrounding area flows into the glass of ice water. Ice melting is an effect, and the equilibrium of temperature is the result. With entropy, local temperatures can be temporarily maintained, but over time, the average temperature of the Universe becomes colder and colder.

The Thinker: But can heat ever lose the battle? There is so much heat in the Universe; there are so many stars in the Universe!

Companion: Stars will burn out. Massive stars last only a few million years, and then they explode to create the second generation of stars; and those new stars, like some others, are likely to run out of fuel in a few billion years. Most stars are not massive enough to become supernovae, and at the end of their life, they do not create new generations of stars. Although many stars last longer than billions of years, they too will exhaust their energy at some point. Yes, there is a lot of heat in the Universe, but it is not enough to keep the Universe warm. In fact, the average temperature of the current Universe has been calculated to be only a few degrees above the absolute zero.

The Thinker: So we keep loosing heat, and everything is getting colder and colder, we are heading toward the point of the temperature of absolute zero!

Companion: There seems to be no other way, even though it might take at least trillions of years to get to that point. During this process, microscopically, the motions of particles within matter become slower and slower, and less and less energy is created to do any work. Then guess what?

The Thinker: What?

Companion: Finally everything stops.

The Thinker: That is it? That is the end? We are going to end up being freezing cold stillness in the dark?

Companion: Ironically, it is called a *heat death*. That is what entropy means in a finite Universe. Entropy increases in one direction and it is irreversible. By the way, that is also how we got the idea of the arrow of time.

The Thinker: It is more like the arrow of the doomed end.

Companion: Sure enough. Derived from the Big Bang Theory, the fate of the Universe has been somewhat calculated, and there are many possible outcomes, including an *open Universe*, a *closed Universe*, and a *flat Universe*. All depend on the mean density of matter of the Universe.

The Thinker: What is the mean density of matter of the Universe?

Companion: It is the total amount of matter there is in the Universe.

The Thinker: Oh. Then what is an open Universe?

Companion: An open Universe expands forever, because it has low density of matter, that is, it does not have enough matter to pull back the expansion of space.

The Thinker: Let me guess. A closed Universe has high density of matter, and it has plenty matter to stop the expansion of space.

Companion: Right. After trillions of years, or however long it might take, the expansion of space will come to a momentary stop, and a closed Universe will collapse on itself. The collapse is called the *Big Crunch*; a reverse version of the Big Bang.

The Thinker: It sounds terrible. What about a flat Universe?

Companion: A flat Universe has the right amount of matter, and its density of matter is termed the *critical density*. A flat Universe expands like an open Universe in the sense the expansion of space might take forever, but it will eventually come to a halt. Then it stays there, flat and forever.

The Thinker: Fascinating! The fate of the Universe can be speculated in such ways.

Companion: Speculation has no limitation.

The Thinker: In the end, what happens to matter in all those space-expansion-based models of universes?

Companion: As you can easily imagine, in a closed Universe, all matter will be thoroughly crunched in the end, thus, the Big Crunch. In a flat Universe, matter will at the end reach a state of maximal entropy, therefore, a heat death. And in an open Universe, entropy also wins. Worse yet, based on the speculation that Dark energy is accelerating the expansion of space at an ever increasing speed, surpass the speed of light at some point, and at the end, all matter will be torn completely apart. And it is called the *Big Rip*. If you wonder why matter is the one to get crunched, frozen, and torn, it is because matter is the one to be actually viewed, touched, and measured.

The Thinker: And what about space? Is there no end to space?

Companion: Well, if space is to be a part of *our* Universe, it has to be part of the fate of *our* Universe, too. If space is like stretchable fabric, it cannot outlast the Universe itself. You can picture that at some point of the expansion, when the limit of the Universe has reached, space will burst open, somewhat like an overstretched balloon, and you can call it the *Big Pop*.

The Thinker: Great!! The Big Crunch, the Big Rip, and the Big Pop, all those big things, is there no other way to get around those deadly outcomes?

Companion: Deadly outcomes are inevitable in a Universe with a pre-set amount of rigid matter and an arbitrary choice of stretchable space. When matter is *set* to be finite, it is bound to be trapped by the inevitability of entropy. When space is *chosen* to be flexible, it serves as a universal benchmark and it becomes our entrapment.

The Thinker: So in a finite Universe we are trapped by both the rigidity of matter and the flexibility of space.

Companion: We have no way to escape. For example; another model of the Universe we have envisioned is called a *cyclic Universe*, which expands and contracts continuously, kind of like a combination of endless open and closed universes. In this model, space appears to be flexible enough to go on forever, but of course, the matter within it is still too rigid to avoid the fatal consequences of entropy. In our finite Universe, we are infinitely trapped by our own rigidity and flexibility.

The Thinker: OK. It seems clear to me that, sooner or later, a catastrophic end is a likely result in a finite Universe. But what does entropy mean in an infinite Universe? What does the end mean in an infinite Universe?

Companion: Nothing, absolutely nothing.

The Thinker: Nothing at all?

Companion: A Universe that is not bound by space, not limited by time, is not determined by any amount of matter. Any finite calculation of increase or decrease of entropy has no effect on the infinity of the Universe.

The Thinker: It is hard for us to imagine and accept infinity!

Companion: But for some of us, it is not too hard to accept the Universe as a *being Universe*, which always exists, requires no creation, and has no fate of the end.

The Thinker: Phew! Then there is no freezing cold end. There is simply no end.

Companion: Whenever we contemplate our origin, we ponder upon our destination. To us, the end is never too far from the beginning. Regarding the Universe, as far as our day is concerned, we might begin it in the morning and end it at night. The day seems to be over, but this is not the end. Not yet.

9
Matter

Matter is what we know; or so we thought.

The Thinker: If the Universe does not begin with the beginning and it does not end with the end; what are we to make of the Universe?

Companion: To make some sense of the Universe, beyond the beginning and the end, let us get to the subject of matter.

The Thinker: Finally!

Companion: Actually, we have been talking about matter all along. When space was presented, we had to present the reference of matter. When time was introduced, we had to introduce the motions of matter. And when we considered our perceptions of space, time, light, temperature, and even our senses, we had to consider them all through matter. So matter is everywhere, and we really want to know what matter is.

The Thinker: Absolutely!

Companion: What comes to your mind when we talk about matter?

The Thinker: Matter is everything. Matter is the composition of physical substance and objects. It is the makeup of actual things; things we can see, touch, and smell; things we can measure, weigh, and compare.

Companion: And remember that all physical objects have *mass*.

The Thinker: Right, and therefore they have *weight*.

Companion: Well, mass and weight are not the same thing.

The Thinker: They are not? A rock weighs more than a sponge of the same size, because the rock has more mass than the sponge. Is it not true that the more mass an object has, the heavier the object is?

Companion: It is true that a more massive object weighs more, but the intuitive connection we make between mass and weight is earthbound. A rock has the same mass on both Earth and Mars, but it weighs differently on the two planets. In other words, mass is the total quantity of matter, and it is the same everywhere in the Universe. Weight is the effect gravity has on matter, and it is different in variant locations in the Universe.

The Thinker: So that means, for matter, mass is constant, but weight is not.

Companion: Right. Do you notice that there is something else constant about matter?

The Thinker: It exists, and it is real?

Companion: Almost. We see matter in various forms, but one thing about matter is that it can be neither created nor destroyed.

The Thinker: How so?

Companion: Matter exists in different phases, but its essence is the same. A familiar example is H_2O, or water. Water is a *liquid* in room temperature; it turns into a *solid* below 0°C; and it becomes a *gas* when the temperature is over 100°C. You see, matter appears in different forms, but the physical elements of matter never disappear, and the volume of matter stays constant for the Universe.

The Thinker: If matter is constant, and matter is the Universe as we know it, does it indicate that the Universe can be neither created nor destroyed?

Companion: It indicates that we do not know matter; and we do not know the Universe.

The Thinker: I think we know a lot about matter.

Companion: Do we really? What is matter made of?

The Thinker: Matter is made of molecules and atoms.

Companion: What are atoms composed of?

The Thinker: Atoms are composed of subatomic particles like protons, neutrons, and electrons.

Companion: What makes up those subatomic particles?

The thinker: Protons and neutrons are made up with elementary particles called quarks. Electrons seem to be just electrons, maybe because they are already much smaller than the protons and neutrons that make up the nuclei of atoms.

Companion: What do elementary particles and electrons consist of?

The Thinker: Elementary particles do not appear to have substructure; and therefore, they are considered to be the fundamental constituents of nature.

Companion: What does fundamental mean?

The Thinker: Fundamental means the most basic, the smallest, the original… Well how should I know? But there has to be something fundamental for everything!

Companion: Indeed we want to find out that fundamental something for everything. That is why we have been trying to break physical objects down into smaller and smaller size to see what actually makes up the whole material world.

The Thinker: And we see elementary particles like quarks.

Companion: Well, we do not really see them. In actuality, no one has ever seen an atom, which is considered to be the smallest organized structure known to man. In comparison, elementary particles make atoms *look* gargantuan. The presence of particles is chiefly detected by man-made instruments. Remarkably, the existence of many particles was first predicted by scientists and scientific theories, and later some predictions would be proved true by confirmations from scientific experiments.

The Thinker: That is great. So can we break matter into ever smaller size, like cutting it into infinitely smaller halves?

Companion: In experiments, the best we can do is to operate particle accelerators to smash particles against one another at a speed near the speed of light; in hope to discover the properties and identities of particles, and to find out what lies underneath those particles, by detections from sophisticated detectors, of course.

The Thinker: What have we found out?

Companion: What we are able to find out is questionable. When particles happen to collide with one another, they break into smaller pieces before vanishing. They leave tracks in the particle accelerator's cloud chamber, and their short-lived moments can be recorded by the detectors, but their results are not always clear cut. And often, we can only describe the results from the particle collisions as energy coming out from the smashed particles.

The Thinker: So there is limitation to what we can do and what we can know.

Companion: We know there is limitation to what we can do, but we do not know what we cannot know.

The Thinker: Does it mean we might know more than we could verify and conform?

Companion: It is hard to say if we think we know more than we actually know, or we know more than we actually think we know. In scientific research, many elementary particles are yet to be found to confirm some of the predictions of scientific theories. Graviton, for instance, is the hypothetical particle that is responsible for transmitting the force of gravity. Common sense tells us that the force of gravity is undeniable, although no graviton has ever been discovered. And in theory, fundamental constituents much smaller than the smallest known particles, smaller than we can ever detect, could exist to make up the physical Universe.

The Thinker: How frustrating! We know that there is something, but this something we can never know. We know the Universe simply to be matter, but we do not know matter, and matter is not simple after all.

Companion: Not only is matter not simple, but it also unnecessarily complicates the Universe, because the Universe requires no matter at all.

The Thinker: What? A Universe without matter? Are you insane?

10

Reality

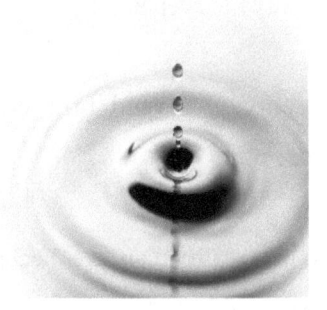

Reality is real, is it not?!

Companion: The Universe requires no matter. We are the ones who demand it.

The Thinker: How can it be! At first, the Universe has no space; that is odd, but the notion is not too hard to comprehend, because space is invisible. Then, the Universe has no time; that is strange, but the concept can be swallowed, after all, time is abstract. And now, the Universe has no matter. What can be said about that? That is ridiculous, and such idea is beyond bizarre. It is totally unacceptable!

Companion: Reality is beyond bizarre whether or not we accept it.

The Thinker: What reality can we accept if we doubt the reality of matter that makes up the familiar world?

Companion: Our acceptance of something has nothing to do with the true nature of that something; and we can never be certain about the familiar world.

The Thinker: Well, right now I am quite certain that we are having a disagreement about matter and reality.

Companion: Disagreements can be agreeable. Let us take a closer look at matter in order to shred some light on reality.

The Thinker: Let us do.

Companion: Let us head to the subatomic realm, since matter is composed of atoms and subatomic elements. Suppose you could shrink yourself to the size of an elementary particle, and you want to find out the reality of an electron.

The Thinker: OK, it sounds like fun. What do I have to do?

Companion: You have to find out where the electron is and how fast it moves; in other words, you need to know the precise location and velocity of that electron.

The Thinker: So I simply measure the speed and mark the location of the electron when I encounter it.

Companion: Yes, but you can never do that. No matter what you do, you can never simultaneously know the precise location and velocity of that electron.

The Thinker: Why not?

Companion: The moment you introduce yourself to the moving electron to determine its location, you inadvertently alter the electron's velocity. In doing so, the more you know about the electron's location, the less you know about its velocity. And if you stay back and let the electron run free in order to determine its velocity, you are not in a good position in locating the electron. In this way, the more you know about the electron's velocity, the less you know about its location.

The Thinker: What kind of scenario is this? Is this a trick?

Companion: No. This is an actual phenomenon from quantum mechanics. Quantum experiments show that in probing subatomic elements, our measurements and observations unavoidably disturb the accuracy of location and velocity of the elements. And without accurate measurements, we have no way of knowing the true reality of anything.

The Thinker: But reality seems to be such a sure thing. Reality still exists with or without our observations, right?

Companion: The only sure thing is that we are not sure. In quantum mechanics, reality is uncertainty and everything is just probability.

The Thinker: What is that again? It sounds confusing.

Companion: Let me use a helium balloon as an analogy to help us to get through the confusion.

The Thinker: OK.

Companion: Imagine that an inflated helium balloon has been put in an empty room for some time. From outside we have no way of knowing where the balloon is located, because the room has no windows and the door is closed.

The Thinker: The balloon is likely to be floating up against the ceiling!

Companion: That is one possibility. It is also possible that the balloon is floating in mid air; it can be up against a corner; it can be down on the floor; it can be somewhere touching the side walls…

The Thinker: OK. You have made your point. The balloon can be anywhere and it has the probability of being at any point in the room. But I still think it has a higher probability of floating up against the ceiling. We can simply open the door to find out where exactly the balloon is, can we not?

Companion: Definitely, we can open the door to see where the balloon is. Once we do that, all the probabilities collapse to one certain outcome. Obviously, the helium balloon is far from quantum mechanics, but the point from it is clear: Prior to observation, it is all probability and nothing is certain about a physical system. The probability of that system becomes a precise quantity only when it is observed.

The Thinker: It seems to me that regardless any probability, our observations still reveal reality, do they not?

Companion: Our observations reveal what we see to be real. There is nothing to be revealed if there is no reality in the first place. It can be said that when we make choice to observe, we actively cause uncertainty into reality.

The Thinker: Now you are saying that reality is subjective. But is it not true that the helium balloon exists inside the room even if we do not open the door to see where it is, and stars still exist even when no one looks up at the night time sky?

Companion: We can maintain our assumption of the dim reality of the helium balloon since we have put it in the room in the first place. But if no one truly looks at the stars to observe their presence, there is no way to tell whether they are there, and it can be argued that the stars do not exist at all.

The Thinker: That is an absurd argument. Why do we not settle this issue objectively? Matter exists, therefore reality exists, end of discussion.

Companion: Not so fast. Let us objectively look at matter again. Remember, we still do not know what fundamental element makes up the Universe, but we know that whatever that element is, it gives matter its quantity, or mass.

The Thinker: That is right. And?

Companion: Mass defines matter, but how does mass come into being?

The Thinker: Maybe mass always exists. It has to exist to make things real.

Companion: Not necessary, many things can exist yet they are massless. Light, for example, is just about as real as everything else, and its particle, the photon, has zero rest mass. It is also why light is generally not considered to be matter. Some other particles, including the hypothetical graviton of gravity, are massless, or are calculated to be massless.

The Thinker: Interesting!

Companion: What is more interesting is that some particles, though being massless, can at some moment and under certain conditions yield mass.

The Thinker: Really? If that is true, it signifies that something massless turns into mass, and in turn mass validates the existence of matter, and then matter consolidates reality. Does it imply that nothing can become something, and we can get something from nothing?

Companion: We can only wonder. Objectively, we are part of matter, and we are in the matter-like reality of our own making. And subjectively, we can make everything into nothing and nothing into everything. Realistically, we create reality.

The Thinker: Wow, this is incredible! We are responsible for the happenings of the Universe. We are matter, reality, and everything. And we are trying to discover, understand, and describe ourselves.

Companion: Incredible and credible is what we are. Without taking ourselves into consideration, how else can we attempt to answer our grandest questions?

Questions including: What is reality? Why do we exist? And above all, why is there a Universe?

The Thinker: Then I cannot help but conclude that the Universe does not require space and time, and it does not require matter and reality. We are the ones who demand them all.

Companion: Any conclusion of any kind is for us to make, just like arrangements of numbers. And in dealing with questions and answers of reality and the Universe, we deal with the existence of numbers.

11
Numbers

Numbers are never outnumbered.

The Thinker: Numbers, how do we get them?

Companion: We invent them.

The Thinker: I know that. But how do they come into existence?

Companion: Due to our need to count, measure, and calculate, we naturally invent many numeral systems. The numeral symbols differ in various systems, but the idea is the same, that is, using symbols to represent quantities of the world of the known and the unknown.

The Thinker: That is quite straight forward. Using numbers, we acquire quantities, such as one apple, two oranges, three bananas, and so on.

Companion: I can tell that you like fruit. But yes, we use numbers to make quantities more comprehensible. And of course, *one*, *two*, and *three* are just words from the English language. The most common numeral symbols we use are *Arabic numerals,* which were invented by Indian mathematicians.

The Thinker: I know. The base ten numbers are *0*, *1*, *2*, *3*, *4*, *5*, *6*, *7*, *8*, and *9*. It is easy enough.

Companion: The system seems to be simple, but its invention is a great human achievement. Based on these ten numbers, which are finite, we can express infinite numbers, however large the numbers might be.

The Thinker: This strikes me as how a numeral system can reflect the Universe. It comes from nothing; it turns into something; and it becomes everything.

Companion: Now you are thinking. Still, in order to make a system work universally, the symbols and meanings of the system have to be universally agreed upon and accepted.

The Thinker: I think the ten Arabic numbers from *0* to *9* are agreeable and acceptable, at least on planet Earth.

Companion: They have to be. We might not accept the definitions of apples, oranges, and bananas, because we might view them differently. We can even disagree with the notions of *1* apple, *2* oranges, and *3* bananas, since no two apples, no two oranges, and no two bananas are alike in the world. But we must agree on and accept the numbers *1*, *2*, and *3*.

The Thinker: It makes me want to say that we can disagree on the content of the Universe, and we do not have to accept the reality we experience, but we must agree on and accept the Universe the way it is.

Companion: Now you are really thinking. But let us not get carried away. We can learn more from the products of our own inventions, as simple as *1*, *2*, and *3*.

The Thinker: Yes, let us learn more from the numbers.

Companion: It should not be a surprise that we invented the number *1* first. Intuitively, the idea of oneness is innate; it represents oneself, a single unity, and the Universe as a whole.

The Thinker: Besides, *1* is the easiest number to count. Well, at least for me. The number *2* came next then.

Companion: There could not have been other way. *2* is not just one more count on top of the *1*; it also represents the two opposite sides of everything we can possibly think of.

The Thinker: Like what?

Companion: Like heaven and Earth, day and night, male and female, and life and death. The number *2* reflects the world so naturally its existence ought to have followed that of the number *1* very quickly.

The thinker: And the number *3* must be next.

Companion: Inevitably. It is one more count on top of the number *2*. However, numbers above *2* were originally counted as *many*. Subsequently, the numbers *3*, *4*, *5*, *6*, *7*, *8*, and *9* followed though the world could be represented with just the first two numbers.

The Thinker: What about the number *0*?

Companion: The realization of the number *0* was a significant leap in abstract thinking for humanity. The concept of zero was best described as *void* and *nothing*. With the invention and application of the number *0*, we no longer have void, and nothing becomes something.

The Thinker: It reminds me again about the infinite Universe. You know; nothing, something, and everything.

Companion: I know what you mean. Numbers are here to stay, and the Universe has widened as we continue to invent more numbers. Not only do we have a number for nothing, but also we have numbers that are less than nothing. We call them *negative numbers*, such as -1, -2, and -3. These numbers are opposite of *1*, *2*, and *3*, which are called *natural numbers*. And so far we have been talking about only *integers*. We have not touched *fractions,* such as *1/2*, *1/3*, and *1/4*, and *decimal* expressions, such as *0.1*, *0.2*, and *0.3*.

The Thinker: I am about to get a headache.

Companion: There are many more types of numbers that exist according to our needs. Some examples are *complex numbers, rational numbers, irrational numbers, real numbers,* and get this, *imaginary numbers.*

The Thinker: I am getting a headache as the Universe widens with our creative ways of coming up with numbers!

Companion: All right, just one more thing about numbers.

The Thinker: What is that?

Companion: Can you see that we are not merely creating a universe with infinite numbers; we are creating infinite universes within infinite numbers?

The Thinker: I think I see the part with infinite numbers. What is the other part about creating infinite universes?

Companion: There are infinite numbers between any two numbers. Pick the numbers *1* and *2*; there are *1.1*, *1.2*, *1.11*, *1.12*… the numbers can continue forever and still maintain within the realm between *1* and *2*. You see, *1* is a finite number, so is *2*. But the membership of the set of the two numbers is infinite.

The Thinker: Now I see it. There is a whole universe between *1* and *2*; there is a whole universe between any of the infinite numbers; and there is everything between everything. The Universe, being greater than numbers and everything, can be nothing else but infinite infinity.

Companion: In spite of infinite numbers, infinite possibilities, and infinite calculations, infinity is not what we usually look for in mathematics. Considering the fact that mathematics is our best ruler in tailoring the Universe; let us take a quick look at math.

The Thinker: No, no math!

12

Mathematics

Calculations are calculated.

But equations are not equal.

Companion: Yes math; and math is inspiring.

The Thinker: Inspiring? I am not fond of math. Its special notation and technical jargon are boring, and all its fancy theorems and complicated equations can make my head spin.

Companion: Believe it or not, although math has become quite difficult for most people to comprehend, its core purpose is to make the Universe easier for all to understand.

The Thinker: For real, by what means?

Companion: By describing. The more accurate we are able to describe the Universe the more likely we can grasp the depth of it. For example, we can better understand the Earth's place in the Universe when we are able to describe the Earth, traveling at 66,000 miles per hour, as the third planet from the Sun that is about 93 million miles away, and that the Earth is one of eight planets that all orbit the Sun, and the Sun, at 500,000 miles per hour, orbits the Milky Way, which also moves at an

	astonishing speed among billions of other galaxies.
The Thinker:	Ah, the Earth is like a miniscule and insignificant dancer in an immense ballroom! It is like painting a picture; only it is done with numbers and math.
Companion:	Painting an actual picture also involves numbers and math. But yes, math is like picture paintings. And to many mathematicians, math is more elegant than any beautiful pictures that anyone can possibly paint.
The Thinker:	No way!
Companion:	You had better believe it. Math evolves from counting, calculation, and measurement. It continues to extend to various branches of applied mathematics with practical applications. At the same time it also develops into pure mathematics, which is just for the benefit of math itself, kind of like art.
The Thinker:	And art is beautiful; in the mind of the beholder.

Companion: Behold, math is considered to be elegant and beautiful not because of its comparison to art but because it provides us with great power in searching for the truth of the Universe. Despite its apparent complexity, math is essentially simple and helpful. We use math to describe phenomena that are too difficult for us to see, experience, and understand. Even when there is nothing for us to see, experience, and understand; we can still have mathematical expressions.

The Thinker: I see! Is that why math is said to be the universal language?

Companion: It is one universal language.

The Thinker: Having equipped with such a powerful tool like math, are we getting close in finding the truth of the Universe?

Companion: Closer than ever but probably will never be close enough.

The Thinker: Please explain.

Companion: On one hand, digging into the Universe with math as our tool, we discover a more and more detailed picture of the Universe. On the other hand, getting stuck with the tools we use, we create bigger and bigger holes in completing the picture of the Universe.

The Thinker: I do not quite get it.

Companion: One thing we like about math is that it draws necessary conclusions, even if the conclusions might have nothing to do with how the Universe works. For example, the equation *1+1=2* satisfies our need for a conclusion, but it reveals almost nothing about the Universe. Technically, mathematical frameworks provide us with indispensable guidelines in painting detailed pictures of the Universe; nevertheless, the pictures we are painting are solely for our own appreciation.

The Thinker: I still need more explanation.

Companion: Let us look at some calculations.

The Thinker: Do we have to?

Companion: Just a few simple equations from Arithmetic, which is a basic branch of mathematics.

The Thinker: Alright.

Companion: Let us pick four random equations:

$$321+57+90+(-468)=0$$
$$50-21-3-4-(89-67)=0$$
$$12345\times0\times6789=0$$
$$0\div12345\div6789=0$$

What do you see in these equations?

The Thinker: I see addition, subtraction, multiplication, and division; I see all ten single digit numbers in each calculation; and I see zero as the answer for all four calculations.

Companion: Correct. What else do you see?

The Thinker: I see there are different ways of coming up with the same result.

Companion: Right. What more do you see?

The Thinker: More? Mmmm. They are mathematically sound equations?

Companion: Sure they are. Let us look at four more equations:

$$1 \times 1 = 1$$
$$1 \div 1 = 1$$
$$1 + 1 = 1$$
$$1 - 1 = 1$$

Now what do you see?

The Thinker: Now I see multiplication, division, addition, and subtraction; I see the number *1* as the only number throughout all these calculations; I also see two true equations and two false equations.

Companion: Correct again. But let us take another look at the four equations; by bringing a pile of sand into our math consideration.

The Thinker: A pile of sand?

Companion: Yes, let the number *1* stand for a pile of sand. Now what do we get when we multiply a pile of sand by one?

The Thinker: We get one pile of sand, since *1×1=1*.

Companion: Right. What do we get when we divide a pile of sand by one?

The Thinker: We get one pile of sand, because *1÷1=1*.

Companion: OK. What do we get when we add a pile of sand to another pile of sand?

The Thinker: We get… one pile of sand?

Companion: Yes, hence *1+1=1*. And what do we get, from this same pile of sand; we take out the pile of sand that we have just added to another pile of sand?

The Thinker: We still get one pile of sand!

Companion: Right, and we have *1−1=1*.

The Thinker: Wait a minute, it must be a trick. Something is wrong here.

Companion: Right again. The numbers can be wrong; the equations can be wrong; or we are the ones to be wrong. In any case, what do you see in the eight equations above?

The Thinker: I see manipulation and arbitration in making up these equations; I see we can make nothing into something and something into nothing; and I see no meaningful connection between what can be made up and what reality might be.

Companion: Bingo! We can always derive some sort of result from our calculations, because our mathematical presentations and possibilities are virtually infinite. Ironically, infinity is not the answer we seek from mathematics.

The Thinker: What do we seek?

Companion: We seek meaning. Oddly, though being ubiquitous, infinity as a mathematical result does not satisfy our need for conclusive answers, and therefore it is considered meaningless. We cannot and we do not want to deal with an infinite Universe simply because we cannot calculate infinity.

The Thinker: So we create infinity only to turn away from our own creation, because it provides no meaning.

Companion: And meaning is something we must have…

The Thinker: Let me finish what you want to say: The Universe requires no math, we are the ones who demand it; and the Universe requires no meaning, we are the ones who demand it.

Companion: Fair enough. Making demands is us, and making us is life.

13
Life

Life is to die for.

The Thinker: Life; what is life?

Companion: To ask what life is, is similar to asking what the Universe is. In observing the presence of the Universe and experiencing the existence of life, we learn and we get to know a lot about life, but at the same time, we do not know enough about it.

The Thinker: What do we know about life, at least life on Earth?

Companion: We know the Earth is teeming with life-forms, including *animals, plants, fungi, protists, archaea,* and *bacteria.*

The Thinker: Some life-forms are more familiar than others.

Companion: Life comes in so many forms we have to organize them into different categories so we can recognize them more easily and more clearly. And we classify every life-form on Earth into *Domain, Kingdom, Phylum, Class, Order, Family, Genus,* and *Species.*

The Thinker: So many names in classification! They do not appear to be easy and clear.

Companion: Let us simplify it by looking at an example of one particular life-form. How about the classification of *human?*

The Thinker: OK. Let us try human. What category is human?

Companion: Humans are in the Domain of *Eukaryota*, which are multi-celled organisms with complex structures.

The Thinker: Humans are definitely complex. What is next?

Companion: Humans belong to the Kingdom of *Animalia,* which are animate eukaryotic organisms.

The Thinker: Humans are animals.

Companion: Humans fit in the Phylum of *Chordata,* which include animals that have vertebrates.

The Thinker: It is true that humans have vertebrates.

Companion: Humans fall into the Class of *Mammalia*, which are mammals.

The Thinker: Humans are also mammals.

Companion: Humans are in the Order of *Primates*, such as monkeys.

The Thinker: Humans are related to monkeys.

Companion: Humans come from the Family of *Great apes*, such as chimpanzees, gorillas, and orangutans.

The Thinker: Humans descend from great apes.

Companion: Humans are from the Genus of *Homo*, which means "man."

The Thinker: Indeed humans are man in the sense of "person."

Companion: And humans are the Species of *Homo sapiens*, which literally means "wise man."

The Thinker: At last, humans have become smart human beings.

Companion: Smart is as smart does. Modern humans further classify themselves into the subspecies of *Homo sapiens sapiens,* which denotes "very wise man."

The Thinker: Is it wise to think that humans are different from other life-forms on Earth because, by far, humans possess the most complexity?

Companion: Humans are the most complex life-form known to man. But it does not matter what humans think, all life-forms on Earth are organisms with carbon-based molecules, and they stay alive through the very chemical processes of life, including continuously replacing molecules with new organic material and emitting old waste.

The Thinker: I suppose that life did not suddenly pop up everywhere on Earth. It must have taken a very long period of time for life to evolve into such complexity and diversity we see today.

Companion: There is indication that simple microbes existed on Earth as early as almost four billion years ago. Initially, it took three billion years for primordial life to evolve into small organisms. Subsequently, from the sea to the land and to the sky, life has evolved and diversified at an ever increasing speed. More and more life-forms have evolved in the process; many millions of species have lived and died. And by a few millions years ago, a very recent moment on the time scale of the biological evolution, the transitions from great apes to humans were under way; and in the last few hundred thousand years, the transformation of modern humans was well in place.

The Thinker: Billions of years of evolution, and life has been around nearly as long as the Earth itself. Why does life thrive in such abundance on Earth?

Companion: One answer is that the Earth is habitable, and there are numerous factors contributing to the habitability of planet Earth. First of all, the Sun is a sustainable star and its energy lasts long enough to allow life to evolve on Earth. Secondly, the Earth is at the right distance from the Sun, and its liquid water neither permanently freezes nor evaporates, making it a suitable environment for life. Thirdly, the Earth holds a substantial atmosphere, a necessary condition for processes of carbon-based organic chemistry and biochemical complexity. Moreover...

The Thinker: OK. Planet Earth has what it takes to develop and sustain life. But how did life begin in the first place? How did nonliving matter become living matter?

Companion: Elements such as carbon, hydrogen, oxygen, and nitrogen necessary for forming building blocks of life are abundant on Earth as well as everywhere else in the Universe; so it is natural to conclude that under the right conditions the emergence of life is but an inevitable result. And undoubtedly life has happened on Earth. However, the very step from nonliving organic chemicals to living organic chemicals is not clear. It seems there has to be *a jump* from nonliving matter to living matter, a jump from nothing to something, almost like the emergence of worldly contents of the Universe; existence miraculously coming out from nonexistence.

The Thinker: So, in other words, we know about life as much as we know about the Universe. We are cognitively observing the existence of life and the Universe, but we do not know how the existence exists. It is so frustrating not to know!

Companion: At least we know why our frustrations exist!

The Thinker: Seriously, if life is an inevitable result from the evolution of the Universe, and it has happened on Earth, can we say that life ought to be elsewhere in the Universe, too?

Companion: It is sensible for us to consider that life can exist on Earth-like planets and moons, although any potential extraterrestrial life-forms are unlikely to resemble that on Earth.

The Thinker: Given the innumerable stars in the Universe, even just a tiny fraction of stars have planets, and only a very tiny fraction of those planets are Earth-like, potential worlds with life have to be in the millions, if not in the billions.

Companion: And given the infinity of the Universe, we are compelled to contemplate the possibility of innumerable worlds of life.

The Thinker: Considering the likelihood that the Universe ought to be teeming with life, it is unavoidable to think that, sooner or later, life-forms from different planets will interact with one another, will they not?

Companion: Maybe sooner or later, but it is easier thought than done. Despite the very idea that countless life-forms might inhabit all over the Universe, the possibility for interplanetary life-forms to come into contact with one another is very remote.

The Thinker: By remote do you mean the distance is too big among the habitable planets?

Companion: That is one seemingly insurmountable obstacle. They can be thousands of light-years or millions of light-years apart. Thousands of years and millions of years is how long it takes just to communicate at the speed of light.

The Thinker: And the life-forms have to be at least as advanced as earthlings to harness the means of radio communication.

Companion: Right, that is another factor. It is reasonable to say that most life-forms might never evolve into intelligent beings and they are not likely to develop into technological societies. On planet Earth, life has diversified into millions of species in processes of billions of years of biological evolution; and human is the only one to make to the dawn of the technological age.

The Thinker: Well, perhaps planet Earth is one of a kind, and mankind is one in millions.

Companion: Being unique or not, for mankind on Earth, the nearest equivalent civilization might be as far as many hundred millions of light-years away.

The Thinker: So for earthlings, close encounters with intelligent extraterrestrial life-forms is a near improbability.

Companion: No doubt distance is a huge obstacle but at least it can be calculated, and given enough time and enough life-forms, the development of civilizations is almost certain to happen. What is uncertain is the sustainability of the civilizations. It is doubtful that any of the developed civilizations can last over hundreds of millions of years for the time needed to make contact with others. They might be destroyed either by nature or by themselves long before any interplanetary meetings can take place.

The Thinker: What a Universe we live in! I am seeing a sad spectacle that countless worlds of life, each one a lonely island and forever separated, living and dying on its own. Can the Universe be so cruel to its inhabitants?

Companion: The Universe generously offers everything, but we are the ones to ignorantly impose cruelty on ourselves. Life-forms, aside from their complexity, are no different from planets, stars, galaxies, and the Universe as a whole. Every part of the Universe is like every cell of the human body; they might exist at different times and at different locations, but they are never isolated. The body connects all the cells, and the Universe connects us all. We are never alone. We can never be alone.

The Thinker: That is a comforting thought, but I still cannot get over my sad feelings. What is it about sentient beings, wanting to know so much about life and the Universe, even though knowledge might not be the final answer to the initial quest? Knowing how life comes to be will not diminish the miracle of life; knowing how the Universe functions will not take away the awesomeness from the Universe; and knowing what we know will not satisfy our feelings about what we know. Why are we so intelligent in learning so much about the Universe when, inadequately, we know so little about how we are supposed to feel?

Companion: What we feel does not depend on what we know. To feel is life and to know is intelligence. Maybe sentient beings are fated in a race between life and intelligence, in which life is always ahead of intelligence.

The Thinker: Life on Earth has come a long way. How far has intelligence come?

14

Intelligence

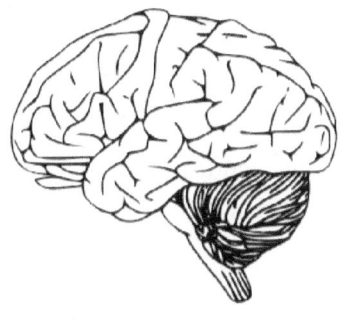

We're smart, don't we think?

Companion: On Earth, intelligence has accompanied life from almost the beginning. For example, unicellular organisms can at least react to tactile stimuli, and early life-forms clung to the environment in the best ways possible in obtaining necessary energy to sustain and to grow, long before they evolved into species that could swim, run, and fly.

The Thinker: Is this just instinct of life, like animal instinct?

Companion: Biological instinct is the most fundamental and crucial intelligence of life. Life would not last without this innate intelligence.

The Thinker: Intelligence has a lot to do with brains, does it not?

Companion: Yes, the brain takes rudimentary intelligence to a higher level, and brainy earthlings commonly attribute intelligence to the size of the brain.

The Thinker: The bigger the brain, the higher the intelligence, is it not?

Companion: Generally it is. In ratio to body weight, humans have the largest brains among all earthly species. Plus, the human brain is the most developed and complex organ known to man. It is an information-processing organ, weighs a few pounds, contains some 100 billion neurons, each connects to 10 thousand others. The complexity of the human brain makes it the most sophisticated control center; and it is utterly responsible for the intelligence that humans possess.

The Thinker: So what exactly is intelligence?

Companion: Although intelligent we might be, we cannot pin point out what exactly intelligence is. From the brain, we acquire a great mental capability, such as to learn, to create, to think logically, to think abstractly, to comprehend ideas, to plan, and to solve problems. You can say that intelligence is power, and the superior brain power gives humans the dominance over the other species on Earth.

The Thinker: Of the essence of the brain, I see thinking as the pinnacle of human intelligence. Only when we are capable of thinking can we harness our brain power and develop a great volume of skills.

Companion: Thinking creatures are what humans are. Being able to think about the Universe and reflect it on oneself is how far intelligence has come on Earth. It also sets humans apart from the other species.

The Thinker: Clearly, humans have no intellectual counterpart on Earth. What about extraterrestrial intelligence? Are there intelligent beings somewhere far away from the Earth?

Companion: It would be very unintelligent for man to think of himself as being in a position of the sole owner of intelligence of the Universe. Intelligently, one needs to learn from oneself before attempting to understand others. By looking at planet Earth, it illustrates that when life exists, intelligence follows, and countless worlds with life in the Universe will mean countless life-forms with intelligence.

The Thinker: So on planets far away from one another, where life might exist, there could be life-forms as intelligent as jellyfish; life-forms as smart as dogs; and life-forms as wise as humans.

Companion: And there could be life-forms far more intelligent, smarter, and wiser than humans.

The Thinker: We cannot help but marvel at the potential of extraterrestrial intelligence, which earthlings have yet to explore. What about artificial intelligence that humans can actually put hands on, such as computers and robots? What can we expect of *AI?*

Companion: AI, being man's creation, is only in its infancy. But in executing certain tasks, AI can already out-perform humans. And AI can even beat the best of human in chess matches. However, it is still a very long way, if it is ever possible, to achieve *strong AI,* which is artificial intelligence that equals to or exceeds human intelligence.

The Thinker: I have no doubt AI will get stronger and stronger in the future, but I just do not foresee any man-made machinery being equivalent to humans. How can machines ever have human traits, such as self-awareness and consciousness?

Companion: Inconceivable as it might seem, one day earthly technology will possibly reach the point of simulating the brain entirely into both software and hardware, and in principle, such complete copy of the brain should be able to perform all tasks, and it should possess all the traits of the original.

The Thinker: My feeling is that a copy of anything can never match, much less surpass the original. And I am not sure such feeling can ever happen to artificial products.

Companion: Mankind has potential to do amazing things. It will be interesting to see if humans can become intelligent enough to create something that is even more intelligent. The limit of AI is unpredictable, but one thing for sure is that life evolves, brains evolve, and intelligence evolves, and such life force will probably be the ultimate boundary to keep artificial intelligence from reaching the equivalence of intelligent life-forms.

The Thinker: I also feel that the forever-evolving condition of life-forms makes each one of us unique. We are not just different from machines; we are different from one another.

Companion: If life-forms have anything in common, it is that they are all different. Two humans possessing the same intelligence will not think in the same way. Thinking is what really sets humans apart, and that is why you can never know entirely what others think.

The Thinker: Sometimes I do not even know what I think!

Companion: What we think, not just how intelligent we are, determines what we know and what we see, and the ways we see the Universe depend on what we think about the Universe.

The Thinker: What do we see in the Universe and what do we think about it?

Companion: Well, let us look at a simple image to demonstrate what we see and what we think about it:

O

What do we see in this image?

The Thinker: I see the number *zero*, the letter *O*, a *circle*, and a *hoop*.

Companion: I see the *Moon*, the *Sun*, a *planet*, and a *galaxy*.

The Thinker: I also see a *balloon*, a *hole*, and a *tunnel*.

Companion: I see a *ripple*, a *bubble*, and an *electron*.

The thinker: Hey, you are seeing in it more than the eye can see.

Companion: Seeing only by the eye is worse than being blind.

The Thinker: Let me think… I see a finite area *inside* the circle and infinite space *outside* the circle.

Companion: And I see the inside is what we *know* and the outside is what we do *not know*.

The Thinker: I also see what we know is *limited* and what we do not know is *unlimited*.

Companion: I also see the limited is *what we are* and the unlimited is *what we think*.

The Thinker: We can go on forever seeing different things from just one thing!

Companion: Yes, because when we think, one thing can never be the same.

The Thinker: So we cannot know the Universe in the same way since we all think differently?

Companion: The Universe is what we think; we are what we think; and we are as intelligent as we think.

The Thinker: Why is it that we are so powerful in thinking, but at the same time we are so powerless to be helplessly bound by our own thinking?

Companion: Thinking reflectively, we recognize the power of our own intelligence and we seem to harness it intelligently, but we are not intelligent enough to be the master of the intelligence of our own power.

The Thinker: No wonder we do not know exactly what intelligence is.

Companion: We do not know exactly what anything is.

The Thinker: We know so much yet we know so little. What do we know? What do we really know about our knowledge?

15

Knowledge

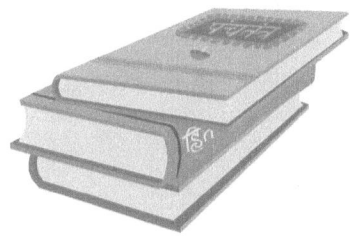

We don't know; that we know.

Companion: Knowledge is what we have been talking about from the very beginning; including things we know and things we do not know.

The Thinker: It is obvious that what-we-know is knowledge; how can what-we-do-not-know be part of our knowledge?

Companion: For all we know; knowing only what-we-know makes the known unknown, and knowing what-we-do-not-know makes the unknown known.

The Thinker: I know we have no way of knowing the unknown. But how can we not know the known?

Companion: Here is something known to us. For a very long time, we had known the Earth to be flat until we figured out that it was a globe. We also had known the Earth to be the center of the Universe until we found out that the Earth orbited the Sun and not the other way around. Then we thought we knew the Sun to be the center of the Universe…

The Thinker: And of course the Sun is not the center of the Universe, because we know that the Sun is just one of the Milky Way galaxy's hundreds of billions of stars, and the Milky Way is only one of billions of galaxies in the Universe. Our world is not flat and we are not the center of the Universe.

Companion: Curious enough, even though modern cosmology has not put a stamp on the shape of the Universe, a flat Universe has been calculated to be highly plausible. And according to the Big Bang Theory, the Universe has emerged from the expansion of a single point and that makes every point in space to be the center of the Universe, and therefore we have no choice but to be the center of the Universe.

The Thinker: Oh nooo! Is that how far we have advanced with such earthly result? Is that how much we have achieved with our scientific knowledge, being self-centered and living in a flat Universe? Is it the best modern science has to offer?

Companion: The best knowledge science can offer us is through the scientific method, which is arguably the most reliable method for acquiring knowledge.

The Thinker: How does the scientific method work?

Companion: The scientific method establishes techniques in systematic ways of collecting data and knowledge by observing, experimenting, and testing hypotheses, which are proposed explanations for phenomena of the Universe. And by applying the scientific method, we continuously increase and refine our knowledge of the Universe.

The Thinker: So we tirelessly observe how the Universe displays, reason why it works, and conclude what it is. The scientific method is not much different from other ways of discovering the Universe.

Companion: When it comes to knowledge, all methods are essentially the same, and the goal is seeking answers to the unknown, although different ways of discovering the same Universe might lead to different answers. Nevertheless, for all our finite knowledge, we need to apply all the possible ways to obtain answers of the infinite Universe.

The Thinker: If the Universe is infinite and our knowledge is finite, we can never have complete knowledge about the Universe. Can we ever get it right about the Universe with our knowledge?

Companion: We do not know if we can get anything right about the Universe, but we know we are getting it through experience, education, and discoveries. The Universe is where our knowledge leads us to.

The Thinker: Where are we now with our knowledge?

Companion: Our knowledge has led us to contemplate the conditions of the Universe; from previous billions of years to trillions of years to come, from possible nothing to apparent everything, and from the seemingly known to the mysterious unknown.

The Thinker: Then we must know a great deal or we would not be able to talk about so many things about the Universe.

Companion: We are capable of talking about as many things as we can imagine, but it does not necessarily mean that we know anything.

The Thinker: What about the subjects we have been discussing? All the questions and answers we have drawn, does it not indicate that we know something?

Companion: We will always have questions because we do not know. We think we know when we come up with some answers only too often to realize later on that we still do not know.

The Thinker: I am not sure what you are talking about since you seem to speak in riddles. Regardless how incomplete our knowledge might be, it ceaselessly accumulates, and we should at least know that we know more and more about the Universe. What I want to know is if our

	knowledge will ever catch up with the infinity of the Universe?
Companion:	We have been trying incessantly to catch up with the Universe. We are in a game of chasing one's own shadow, forever increasing is the nature of knowledge, and so is the majesty of the Universe.
The Thinker:	But we are unstoppable! As we continue to make great scientific discoveries, all of a sudden, the Earth has become a very small world. We are no longer bound by a tiny blue speck. We exceed the limit of our sky and we are reaching out toward the whole Universe.
Companion:	We could discover everything, and in doing so possibly diminishing the immensity of the Universe. However, our ultimate challenge might not be how much we need to know outwards, but how well we should know inwards. Aside from knowing everything else, can we really know the knower that is us?
The Thinker:	So the Universe is not the limit; we are the limit!

Companion: We are the ones to set the limit with our mind, and why should there be limit to the mind?

16

The Mind

No mind is absent-minded.

The Thinker: What is the mind?

Companion: The mind includes all the traits of intellect, all the aspects of the conscious, the subconscious, and the unconscious, and everything else that might come to mind.

The Thinker: That is mindful! Is the mind the same as the brain, or is it independent from the physical body?

Companion: You are asking the *mind-body problem*, about which many great minds have wondered.

The Thinker: What is the problem?

Companion: Among many views about the mind there are two opposite ones: One sees the mind as equivalent to the physical body, and the other sees the mind to be separate from any physical entity.

The Thinker: Well, my scientific mind sees the mind the same as the brain. All the mental attributes, such as thoughts, beliefs, desires, and perceptions, are identical to neuronal phenomena. After all, there would not be any mind without the physical brain.

Companion: Scientifically, you are likely to be right.

The Thinker: But my non-scientific mind sees the mind to be much more than the brain. How can all the everlasting greatness be constrained in the mortal brain? There is no way that unbound imagination of the infinite Universe can be confined in a finite size of physical matter.

Companion: Non-scientifically, you cannot be wrong.

The Thinker: Which is it? Is the mind physical or non-physical? I cannot possibly be correct being on two totally opposite sides!

Companion: Why not? In the field of science, light is both wavelike and particle-like, and they are two strikingly different properties. Away from science, light can be anything there is. Can the mind be any less than light?

The Thinker: If the brain is just a few pounds of physical matter, how do we measure the volume of the mind? We can break apart every component of a tangible body, but how do we dissect any intangible element of the mind?

Companion: You can say that the mind is greater than the body, because the whole is greater than the sum of its parts.

The Thinker: So limited is the body and so extraordinary is the mind, and we possess both of them!

Companion: We possess all. We possess the Universe; a Universe with a boundary, but the boundary, without limit.

The Thinker: For my bound possession, my body is with the cosmos, which is physical.

Companion: For my unbound possession, my mind is with the Universe, which is transcendent.

The Thinker: And for all there was, is, and will be, my thought
(Companion) is the Universe, which is what we are.

The Thinker is the Companion, and the Companion is the Thinker. They are in one of space, time, and matter; they are in one of one life-form, one planet, and one star; they are in one of questions, answers, and imaginations; and they are in one of the body, the mind, and the Universe. This is everything; this is the beginning; and this is the end.

The end ends where

we stop thinking.

About the author

Feilong Wu is a national and international competitive diving champion from China. He has been a diving coach, a martial art instructor, and a certified personal trainer. Also he has been a performer for Cirque Du Soleil and SeaWorld. His extensive athletic background and accomplishments might seem to have nothing to do with writing ***The Universal Thinker***; however, possessing great physical skills only enhances his mental ability to think beyond the physical aspects of our world. Being an enthusiast of science and a thinker of philosophy, Feilong strikes a balance, scientifically and philosophically, between the two. He currently lives in the US.

Feilong Wu can be reached at: feilongwu3000@yahoo.com

www.ingramcontent.com/pod-product-compliance
Lightning Source LLC
Chambersburg PA
CBHW032021170526
45157CB00002B/793